vjbnf C0-DYA-898
629.892 TROUP

Troupe, Thomas Kingsley, author
Underwater robots
33410015618491   07-13-2021

Porter County
Public Library System
103 Jefferson Street
Valparaiso, IN 46383

Mighty Bots

UNDERWATER ROBOTS

THOMAS KINGSLEY TROUPE

WORLD BOOK

This World Book edition of *Underwater Robots* is published by agreement between Black Rabbit Books and World Book, Inc.
© 2018 Black Rabbit Books,
2140 Howard Dr. West,
North Mankato, MN 56003 U.S.A.
World Book, Inc.,
180 North LaSalle St., Suite 900,
Chicago, IL 60601 U.S.A.

All rights reserved. No part of this book may be reproduced in any form without written permission from the publisher.

Marysa Storm, editor; Grant Gould, interior designer; Michael Sellner, cover designer; Omay Ayres, photo researcher

Library of Congress Control Number: 2016049964

ISBN: 978-0-7166-9334-5

Printed in the United States at CG Book Printers,
North Mankato, Minnesota, 56003. 2/18

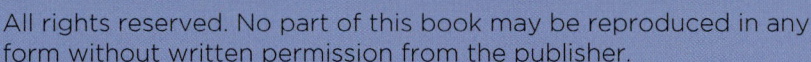

Image Credits

Alamy: Aflo Co. Ltd., 12; Nature Picture Library, 17 (jellyfish); commons.wikimedia.org: Jaan Rebane, 14 (U-CAT); NOAA/Institute for Exploration/University of Rhode Island, 22 (bottom); http://www.ttu.ee/Flickr Stream: mihkel l, 21 (U-CAT); http://www.whoi.edu: Woods Hole Oceanic Institution, 8 (Nereus), 18–19, 22–23 (background), 31; iStock: pichitstocker, 11 (top); liquid-robotics.com: Liquid Robotics, 24; losyziemi.pl: Unknown, 17 (octopus); Newscom: GH1/KIOST, Cover, 15; noaa.gov / Rutgers University, Unknown, 22 (top); Science Source: British Antarctic Survey, 23 (top); Danté Fenolio, 16; Peter Menzel, 14 (robolobster); United States Antarctic Program., 23 (both bottom photos); SeaEYE: SAAB, 28–29; Shutterstock: Angelo Giampiccolo, 21 (ship); Bruce Johnstone, 32; Iakov Filimonov, 1, Back Cover; Levent Konuk, 8–9 (background); Pixone, 11 (bottom); Rich Carey, 28; vanhurck, 3; US Navy / commons.wikimedi.org: Photographer's Mate 2nd Class Dawn C. Montgomery., 4–5; www.nasa.gov: NASA, 27; www.openrov.com: Open ROV, 6

Every effort has been made to contact copyright holders for material reproduced in this book. Any omissions will be rectified in subsequent printings if notice is given to the publisher.

CONTENTS

CHAPTER 1
Underwater Adventures..........4

CHAPTER 2
Underwater Workforce..........10

CHAPTER 3
Underwater Explorers.............16

CHAPTER 4
Record Breakers.....25

Other Resources...........30

CHAPTER 1

UNDERWATER
Adventures

Far from shore, waves rock a boat in the Atlantic Ocean. From the boat, a yellow machine is lowered into the water. It disappears beneath the waves. The underwater robot's **mission** has begun! It is on its way to the deepest, darkest parts of the ocean. There, it will collect samples and take video.

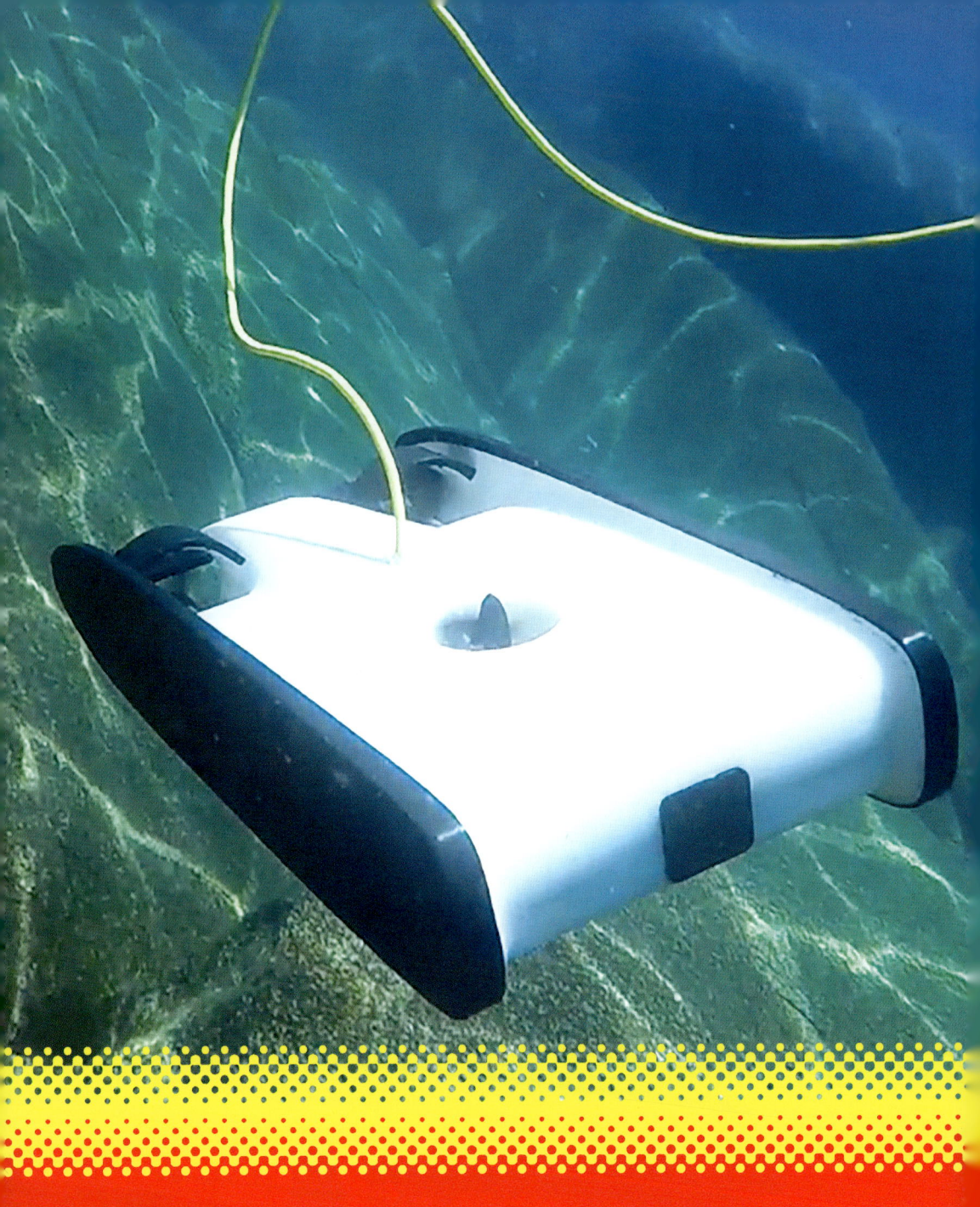

**The Trident is an ROV.
ROV stands for Remotely Operated Vehicle.**

Deep, Deep Dives

Exploring and working in oceans is **risky**. To stay safe, people use robots. Some bots help fix and build. Others discover new places and animals.

Most underwater robots, such as the Trident, are **drones**. Drones are remote-controlled. An operator tells them where to go. Many of these bots have built-in cameras. The cameras send images to screens on the surface. The screens show the users where the bots are going.

COMPARING DIVING DEPTHS

feet below surface

	SNORKELING	RECREATIONAL SCUBA DIVING	TECHNICAL SCUBA DIVING

0
3,000
6,000
9,000
12,000
15,000
18,000
21,000
24,000
27,000
30,000
33,000
36,000

1 foot (.3 meter) people can swim at surface

130 feet (40 m) safe diving depth with wet suit, mask, and air tank

350 feet (107 m) safe diving depth with proper gear and advanced training

NEREUS (ROV)

35,768 feet (10,902 m)
one of the deepest ROV dives

DEEPSEA CHALLENGER

35,787 feet (10,908 m)
deepest solo submarine dive

TRIESTE

35,813 feet (10,916 m)
one of the deepest sub dives

CHAPTER 2

UNDERWATER
Workforce

Some offshore oil platforms use ROVs. Oil platforms are big factories. They pull oil from under the ocean floor. These platforms can be difficult to build. They are also hard to fix. That's why workers use ROVs. These robots help with repairs. They can bury and inspect pipelines. Fewer humans are in danger when robots do the work.

Some ROVs have robotic arms. The arms can move and cut objects.

Inside Pipes

Some places, such as inside pipes, are hard to inspect. But small bots can enter pipes. They can reach what people cannot.

The TRACT project made a snakelike robot. The robot can examine pipes filled with water. **Propellers** move the bot through pipes. Using **ultrasound**, the bot finds where pipe metal is thin. These are the spots that need attention.

U-CAT
SWIMS LIKE A TURTLE

INSPIRED BY NATURE
Some robots look like animals.

ROBOLOBSTER
WALKS LIKE A LOBSTER ACROSS GROUND UNDERWATER

CRABSTER......
MOVES LIKE A CRAB BUT WEIGHS AS MUCH AS A COW

CHAPTER 3

Underwater EXPLORERS

Skilled divers can only reach depths of about 350 feet (107 m). So people use robots to explore. These machines help people learn about the oceans. Without robots, many discoveries wouldn't have been made. Many creatures and places would still be unknown.

Deep-Sea Creatures

ROVs discovered these animals.

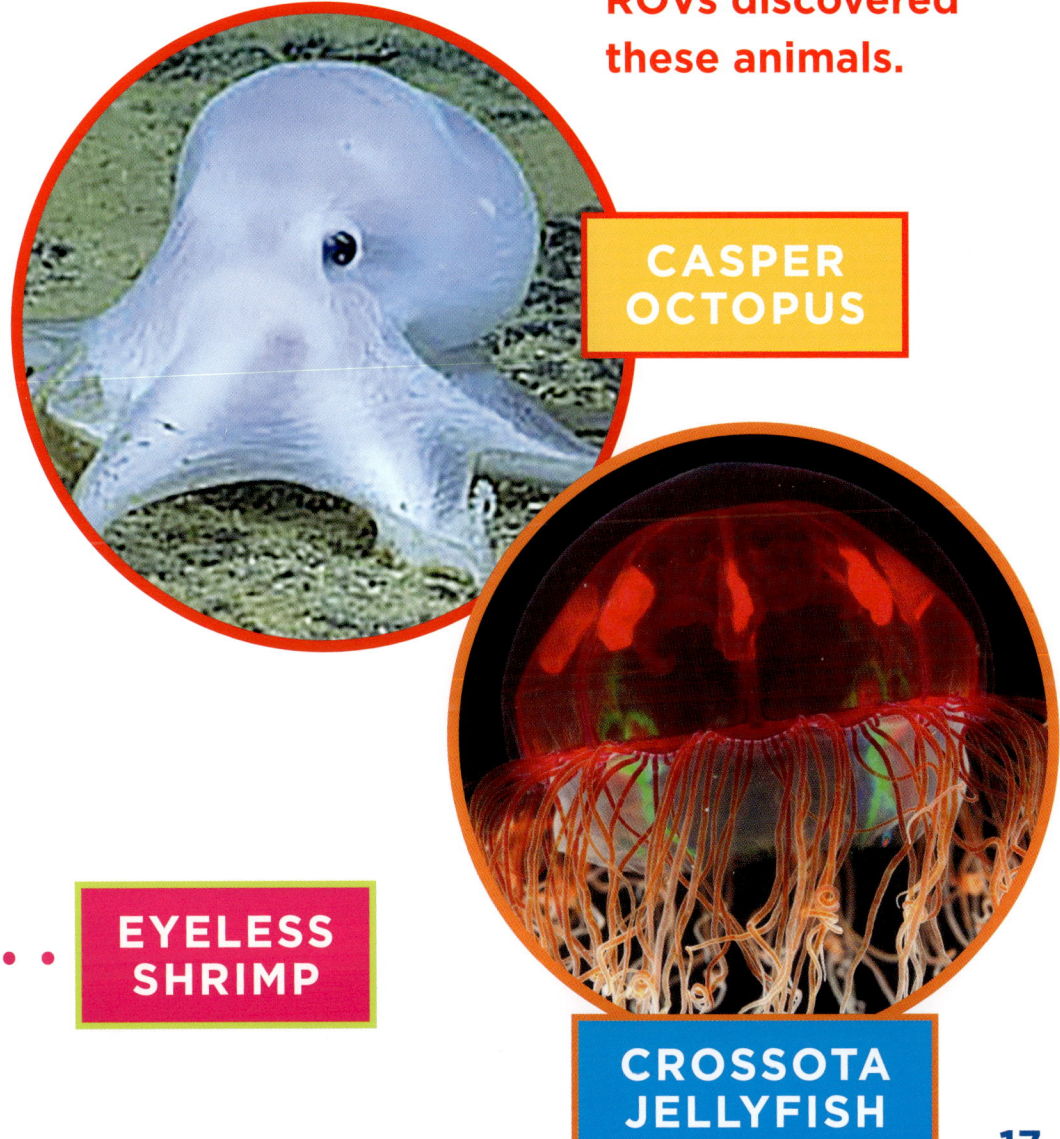

CASPER OCTOPUS

EYELESS SHRIMP

CROSSOTA JELLYFISH

Mapping Machine

The seaBED looks like two **torpedoes**. But the robot isn't used to attack. It's used to make maps. Using sound waves, it makes pictures of ocean floors. The bot mapped the ice in Antarctica.

The seaBED travels at only .6 mile (1 kilometer) per hour.

Shipwreck Swimmer

The U-CAT isn't a big bot. Instead, it's small and simple. It can easily go places people can't. People built the U-CAT to explore shipwrecks. The bot swims smoothly. It doesn't stir up the ocean floor. People hope to use the bot to find items from the past.

• • • • • • • • • • • • • • • • • • • •

The bot can swim without being told to. It uses sound waves to avoid hitting objects. Beacons let the bot know where it is.

EXCITING EXPLORATIONS

Robots explore cool places. They make amazing discoveries.

2009
Atlantic Ocean
Bot crosses the ocean.

1980 |||||||||||||||||||||

1986
Titanic
Bot takes first pictures of the wreck.

22

2010 and 2012
Antarctica
Bot explores underwater areas near Antarctica.

2016

2011
volcanic vents
Bot finds new sea life.

2015
new reef
ROVs discover huge new reef.

23

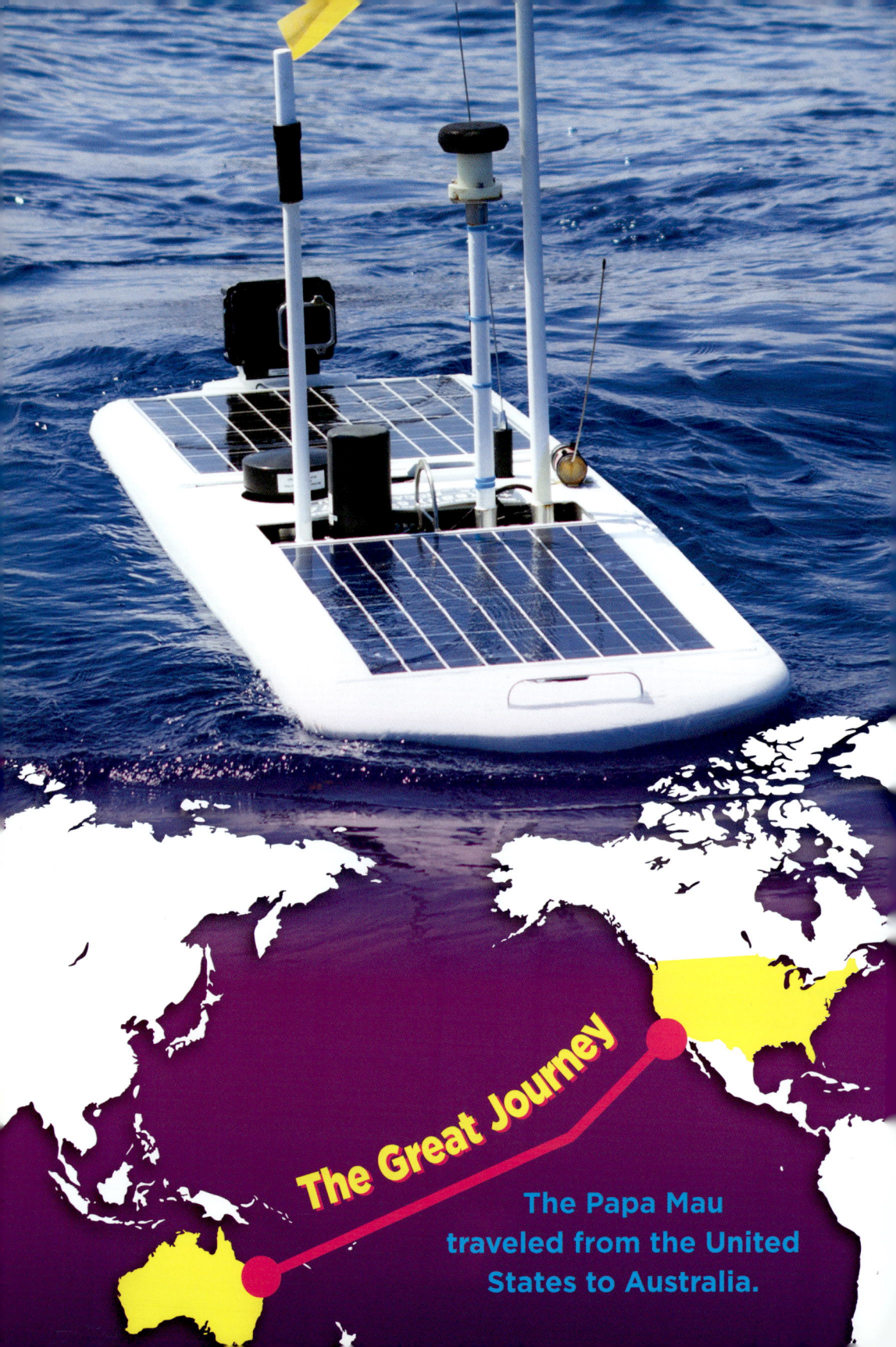

The Great Journey

The Papa Mau traveled from the United States to Australia.

CHAPTER 4

Record
BREAKERS

Robots make and break records. Some travel to new depths. Others go great distances.

From 2011 to 2012, the Papa Mau traveled 9,000 miles (14,484 km). The self-controlled bot traveled using only wave and solar power. It had to survive big waves, sharks, and storms. While it traveled, the bot studied the Pacific Ocean. The Papa Mau proved simple bots could go far.

A Sea in Space?

Exploring seas on Earth is just the beginning. Someday, bots will swim through seas in space. Right now, NASA is working on a robot sub. It might search the sea on Titan. Titan is Saturn's largest moon. The sub will dive into the sea. Then it will gather data. It will measure **chemicals** in the sea. It'll also map the sea. The bot will send the information back to Earth.

The bot is called the Titan Sub.

Into the Deep

Oceans are incredible places. They are just begging to be explored. And robots will continue to help with the exploration. Maybe someday, humans will be able to see it all!

GLOSSARY

beacon (BEE-kuhn)—a radio signal that is broadcast to help guide ships, airplanes, and other machines

chemical (KE-muh-kuhl)—a substance that can cause a change in another substance

drone (DROHN)—an unmanned aircraft or ship guided by remote control or onboard computer

mission (MISH-uhn)—a task or job that someone is given to do

propeller (pruh-PEL-er)—a device with two or more blades that turn quickly and cause a ship or aircraft to move

recreational (rek-ree-EY-shuh-nl)—done for fun or enjoyment

risky (RIS-kee)—involving the possibility of something bad or unpleasant happening

torpedo (tawr-PEE-doh)—a bomb shaped like a tube that is fired underwater

ultrasound (UHL-truh-sound)—a method of producing images by using sound waves that are too high to be heard

LEARN MORE

BOOKS

Faust, Daniel R. *Underwater Robots.* Robots and Robotics. New York: PowerKids Press, 2016.

Spilsbury, Richard, and Louise Spilsbury. *Robots Underwater.* Amazing Robots. New York: Gareth Stevens Publishing, 2016.

Swanson, Jennifer. *National Geographic Kids. Everything Robotics: All the Robotic Photos, Facts, and Fun!* Everything Series. Washington, D.C.: National Geographic, 2016.

WEBSITES

Robotics
kidsahead.com/subjects/1-robotics

Robotics: Facts
idahoptv.org/sciencetrek/topics/robots/facts.cfm

Robots for Kids
www.sciencekids.co.nz/robots.html

INDEX

C
Crabster, 15

D
discoveries, 16–17, 23

H
history, 22–23

P
Papa Mau, 24–25

R
RoboLobster, 14

S
seaBED, 19

T
Titan Sub, 26–27
TRACT project, 13
Trident, 6–7

U
U-CAT, 14, 20